# Animal Clues Guess Who

# POLAR ANIMALS

Kizzi Roberts
M.S. Animal Science

## Ways to use this book

### Younger Readers

Read the book together and let your reader hear the clues and see the close-up picture as they guess the animal.

### Older Readers

Read each clue aloud without showing the close-up picture. Let your reader try to guess the animal as you read the clues. Show the close-up picture if they need an extra clue.

### Make It a Game!

Read the clues aloud and try to guess the animal in as few clues as possible. Come up with your own clues for each animal after reading the extra facts.

LEARNINGSPARKEDUCATIONALPUBLISHING

I eat **LEMMINGS** and sometimes **INSECTS** and **BERRIES**.

I do not need to **HIBERNATE**.

I have **SMALL** ears.

I have a thick **WHITE** coat in **WINTER**.

# WHAT AM I?

# I am an
# ARCTIC FOX.

# ARCTIC FOX FACTS

• • • • • • • • • • • • • • • • • • •

The Arctic fox has a thick, fur coat to keep its body temperature at 104 degrees Fahrenheit even in freezing conditions. When it curls up to sleep, its tail covers its face and nose like a blanket.

The Arctic fox blends into the Arctic environment. Its coat changes color depending on the season. In winter, its white coat blends in with the snow. In summer, its brown or gray coat matches the rocks and dirt.

The Arctic fox mainly eats small rodents called lemmings. The fox hears lemmings in their snow tunnels. It leaps up high and dives into the snow to catch them. When food is hard to find, the fox curls up in a den and sleeps for up to two weeks. The fox doesn't officially hibernate, but sometimes it conserves energy when food is unavailable.

I only eat **PLANTS**.

I live in a **HERD**.

I have large, hollow **HOOVES**.

I have **ANTLERS**.

**WHAT AM I?**

I am a **CARIBOU.**

# CARIBOU FACTS

• • • • • • • • • • • • • • • • • • • • • • •

Caribou eat a variety of plants. They dig their large hooves in the snow to find lichen. Lichen, composed of fungus and algae, is an important winter food source for caribou and other Arctic animals.

Caribou live together in large herds. During summer months, some caribou herds grow to more than ten thousand. Caribou live in many parts of the world. The largest herds are in Alaska, Canada, and Russia.

Both male and female caribou have antlers. In all other deer species, only males have antlers. Female caribou keep their antlers through the winter, while males lose their antlers in late fall. Caribou use their antlers for fighting, defending their young, and guarding food sources.

I am nicknamed the **SEA PARROT**.

I am mostly found in **ICELAND** and **NORWAY**.

I **DIVE** into water to catch my food.

I can **FLY** and **SWIM**.

# WHAT AM I?

I am a **PUFFIN.**

# PUFFIN FACTS

The puffin is known for its brightly colored beak. Due to its beak and living along coastlines and on islands, it has earned the nickname "Parrot of the Sea".

There are three kinds of puffins: the Atlantic puffin, the horned puffin, and the tufted puffin. The horned puffin and the tufted puffin live on land in the Pacific Ocean. The Atlantic puffin is the most common type of puffin. It lives around the Atlantic Ocean. About 90 percent of Atlantic puffins live in Iceland and Norway.

The puffin is good at both flying and swimming. It flies up to 50 miles per hour. It flies while hunting small fish, then dives into the water. Its webbed feet make the puffin a strong swimmer. The puffin catches a lot of fish in its beak. It has special spikes inside its beak that help keep all the fish in its mouth.

I can **RUN** fast.

I only eat **PLANTS**.

I can **HOP** on my back legs.

I have long, sharp
front **TEETH**.

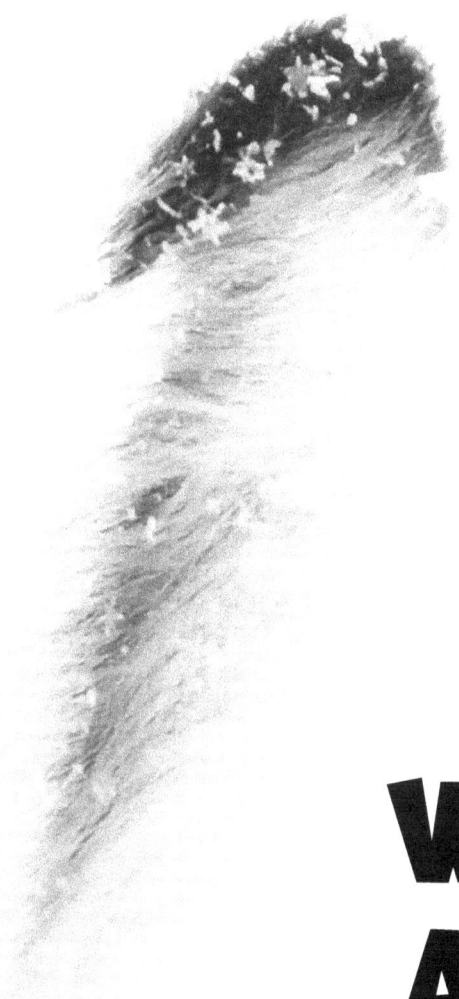

**WHAT AM I?**

I am an **ARCTIC HARE.**

# ARCTIC HARE FACTS

· · · · · · · · · · · · · · · · · · · · ·

The Arctic hare is well adapted to the cold polar region. Even though it is the largest hare in North America, it has much shorter limbs and ears than other hares. The hare's short limbs and ears conserves body heat during cold winter months.

The Arctic hare's thick fur coat keeps it warm. Its coat also protects it from predators by helping the hare blend in with the Arctic environment. In winter, the Arctic hare has a white coat. In warm months, the hare's coat turns brown or gray to blend in with rocks and dirt.

The Arctic hare has strong back legs that help it run fast. It can also hop like a kangaroo! It uses the long claws on its back legs to dig in the snow for food and shelter. It also uses its long, sharp front teeth to pull food from between rocks.

I am the **LARGEST** meat-eating animal on **LAND**.

I can smell food up to **TEN** miles away.

I hunt for **SEALS**.

I have **BLACK** skin.

**WHAT AM I?**

I am a **POLAR BEAR.**

# POLAR BEAR FACTS

· · · · · · · · · · · · · · · · · ·

The Polar bear is a huge animal. It can grow to over 8 feet long and weigh around 1500 pounds. It is the largest carnivore (meat eater) on land, and spends about half its life hunting for food.

The Polar bear spends much of its life on the frozen ocean ice. The Polar bear also swims in the icy ocean. It can swim for hours in the cold water, but it doesn't hunt while swimming. Instead, it waits by holes in the ice when hunting its main food source, seals. It waits for the seals to come up to breathe and then catches them. The Polar bear can smell a seal up to 10 miles away.

The Polar bear's thick fur appears white, but the hairs are actually hollow. Its hair reflects light, making it appear white. The Polar bear has black skin which helps it stay warm by absorbing sunlight.

I drink **SALTWATER** while swimming.

I **ONLY** live in the **SOUTHERN** hemisphere.

I am **EXCELLENT** at swimming.

I can **NOT** fly.

WHAT AM I?

I am a
PENGUIN.

# PENGUIN FACTS

There are 18 different species of penguins, and they all live South of the equator. Only 7 species of penguins live in icy climates such as Antarctica. Most penguin species live in warmer areas, and some even live in tropical climates.

No matter where they live, all penguins flightless. They are, however, excellent swimmers. A penguins swallows a lot of seawater while swimming. Luckily, a special gland filters the salt out of its bloodstream. The salt is expelled through its nose. Penguins shake their heads a lot to remove the salt from their nostrils.

Emperor penguins are the largest species of penguin at almost four feet tall and weighing over 77 pounds. These penguins lay their eggs and raise their young in Antarctica. Their dense feathers protect them from cold weather. Their feathers are more dense than any other bird.

I am the largest member of the **WEASEL FAMILY**.

I need **DEEP SNOW** to make a den for my **BABIES**.

I am known for being **FEROCIOUS**.

I have a strong **ODOR**.

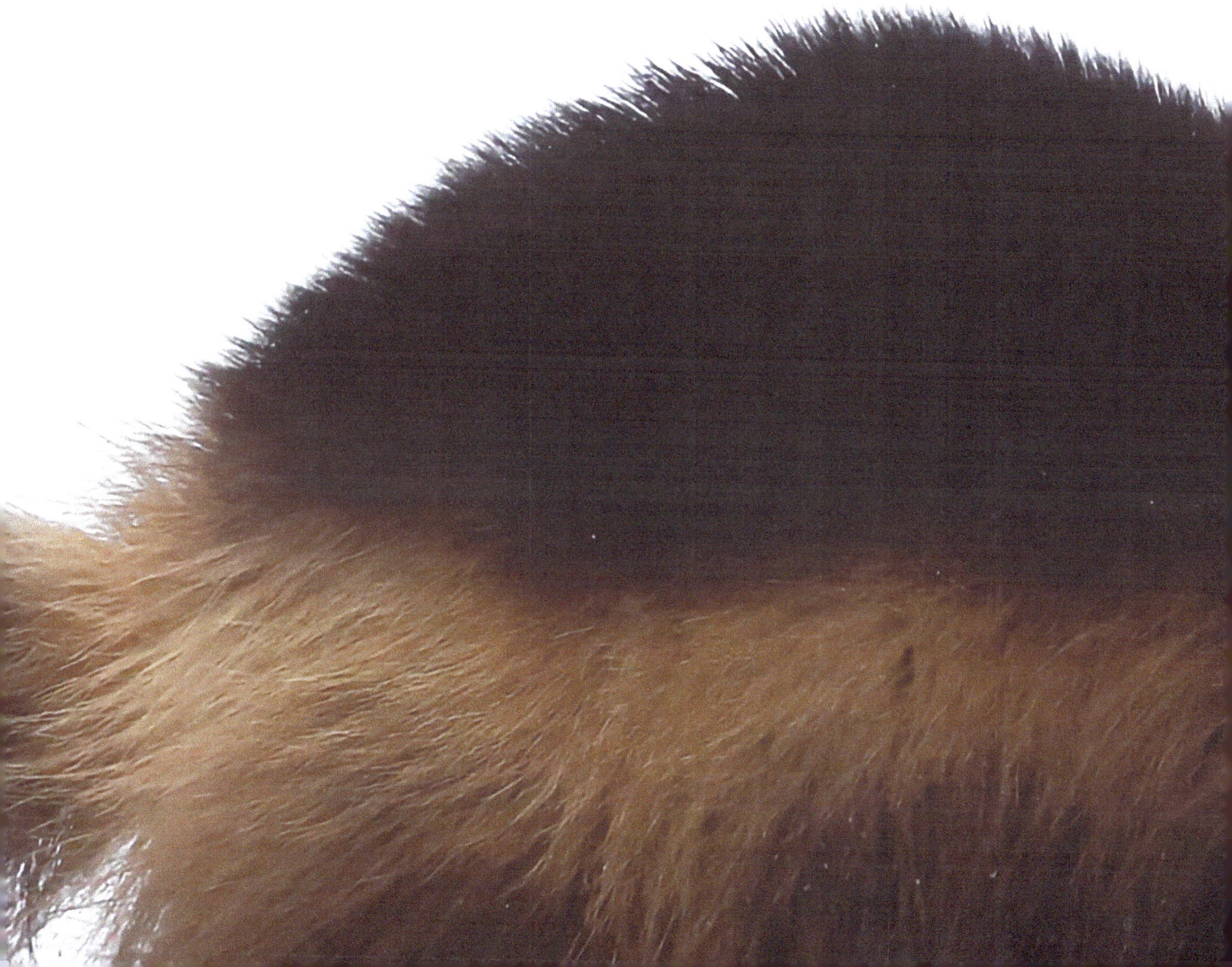

WHAT
AM I?

I am a **WOLVERINE**.

# WOLVERINE FACTS

• • • • • • • • • • • • • • • • • • • •

The wolverine is the largest member of the weasel family, also known as mustelids. Like many mustelids, the wolverine has a strong odor. This odor comes from scent glands it uses for marking its territory and communicating with other wolverines.

The wolverine has many nicknames, such as skunk bear, carcajou, and nasty cat. It has earned these nicknames due to its ferocious behavior when standing up to much larger animals. Despite only weighing between 30 and 50 pounds, wolverines face down moose, bears, and wolves and often win.

The wolverine requires deep snow to make a den for its babies. A female wolverine digs a den at least five feet into the snow. The den protects the babies from the cold weather and from predators. Wolverines are often found high in mountains where deep snow is abundant.

# Clues by YOU!

It's your turn to create clues. Look at the pictures and read the facts for the following Polar animals. Can you come up with four clues for each animal based on their appearance, behavior, or some other fact?

Tell your clues to a friend or family member and see if they can guess the animal without looking at the picture.

## I am an ARCTIC WOLF.

Arctic wolves are also known as Polar wolves or white wolves. Unlike a lot of other polar animals, the Arctic wolf keeps its white fur all year long.

Arctic wolves only eat meat. They hunt large mammals like caribou, and small mammals like hares.

Arctic wolves live in the harshest Arctic environments where the ground is always frozen. They rarely make contact with humans, unlike other types of wolves.

# I am a SNOWY OWL.

The snowy owl spends summers in the Arctic, although sometimes it stays all year. It is the heaviest owl in North America and can weigh up to 6.5 pounds with a 4 to 5 foot wingspan.

Male owls are smaller than females and almost completely white. The snowy owl hunts by day or night for small rodents. Owls swallow small prey whole.

# I am a HARP SEAL.

Harp seals are born covered in completely white fur. Adults have gray fur with black patches.

Harp seals spend almost their entire life in the water. Adults can stay underwater for up to 15 minutes. They leave the water to have their babies on pack ice. Pack ice consists of huge frozen sections of the ocean that are not attached to land.

Photographs © slowmotiongli/depositphotos.com; Photocech/depositphotos.com; JimCumming/depositphotos.com; cybernesco/depositphotos.com; richardsjeremy/depositphotos.com; Alexey_Seafarer/depositphotos.com; peterwey/depositphotos.com; nicholas_dale/depositphotos.com; PantherMediaSeller/depositphotos.com; lightpoet/depositphotos.com; arepiv/depositphotos.com; Alexey_Seafarer/depositphotos.com; jill@ghostbear.org/depositphotos.com; jeffbanke/depositphotos.com; akinshin/depositphotos.com; OndrejProsicky/depositphotos.com

Published in March 2024 by Learning Spark Educational Publishing in Rogersville, Missouri. Learning Spark Educational Publishing is an imprint of Elemental Ink Publishing LLC.

Library of Congress Number: 2024904065

Hardcover: 979-8-88884-025-2; Paperback: 979-8-88884-024-5
Edited by Carrie Rodell. Book design and layout by Kizzi Roberts.

www.LearningSpark.com

www.ingramcontent.com/pod-product-compliance
Lightning Source LLC
Chambersburg PA
CBHW042344030426
42335CB00030B/3452